SPOTLIGHT ON NATURE

SEAL

JULIE AND DAVID LAVENDER

I0895284

CREATIVE EDUCATION · CREATIVE PAPERBACKS

Published by Creative Education and Creative Paperbacks
P.O. Box 227, Mankato, Minnesota 56002
Creative Education and Creative Paperbacks are imprints of The Creative Company
www.thecreativecompany.us

Design by Rhea Magaro; production by Beeline Media & Design, Inc.
Art direction by Tom Morgan
Printed in the United States of America

Photographs by Alamy (Wolfgang Kaehler, Frank Bienewald, Arndt Sven-Erik), Dreamstime (Hakoar, Slowmotiongli), Getty Images (Andrew Peacock, David Madison, by wildestanimal), Shutterstock (Jiang Zhongyan, Xavier Hoenner, Eric Isselee, Arthur Greenberg, Alfmaler, johavel, hachiko.58, RugliG, WitR, COULANGES, Ian Dyball, Lorraine Logan, RugliG, Natalia Golovina)

Copyright © 2025 Creative Education, Creative Paperbacks
International copyright reserved in all countries. No part of this book may be reproduced in any form without written permission from the publisher.

Library of Congress Cataloging-in-Publication Data
Names: Lavender, Julie, author. | Lavender, David R., 1961- author.
Title: Seal / Julie and David Lavender.
Description: Mankato, Minnesota : Creative Education and Creative Paperbacks, [2025] | Series: Spotlight on nature | Includes bibliographical references and index. | Audience: Ages 10-13 | Audience: Grades 4-6 | Summary: "Dive into the seal's world with our new book for mid-level readers. An engaging narrative of a seal family, captivating images, infographics, milestones, and resources make it perfect for young nature enthusiasts"-- Provided by publisher.
Identifiers: LCCN 2023059491 (print) | LCCN 2023059492 (ebook) | ISBN 9798889891963 (library binding) | ISBN 9781682775691 (paperback) | ISBN 9798889892045 (ebook)
Subjects: LCSH: Northern elephant seal--Juvenile literature. | Northern elephant seal--Life cycles--Juvenile literature. | Northern elephant seal--Behavior--Juvenile literature. | Northern elephant seal--California--Point Reyes Peninsula--Juvenile literature. | Elephant seals--Juvenile literature. | Seals (Animals)--Juvenile literature.
Classification: LCC QL737.P64 L378 2025 (print) | LCC QL737.P64 (ebook) |
DDC 599.79/4--dc23/eng/20240125
LC record available at https://lccn.loc.gov/2023059491
LC ebook record available at https://lccn.loc.gov/2023059492

CONTENTS

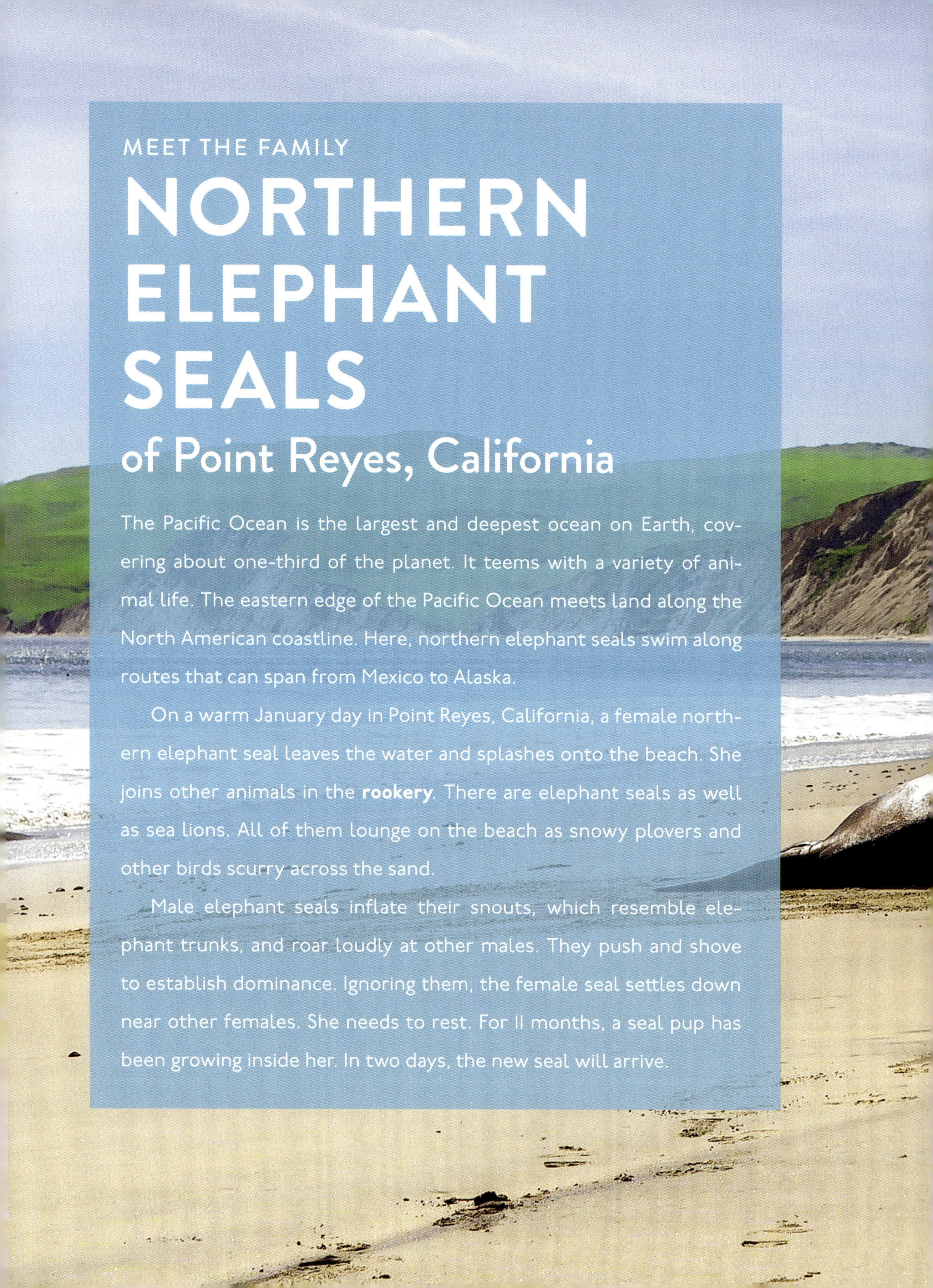

MEET THE FAMILY

NORTHERN ELEPHANT SEALS

of Point Reyes, California

The Pacific Ocean is the largest and deepest ocean on Earth, covering about one-third of the planet. It teems with a variety of animal life. The eastern edge of the Pacific Ocean meets land along the North American coastline. Here, northern elephant seals swim along routes that can span from Mexico to Alaska.

On a warm January day in Point Reyes, California, a female northern elephant seal leaves the water and splashes onto the beach. She joins other animals in the **rookery**. There are elephant seals as well as sea lions. All of them lounge on the beach as snowy plovers and other birds scurry across the sand.

Male elephant seals inflate their snouts, which resemble elephant trunks, and roar loudly at other males. They push and shove to establish dominance. Ignoring them, the female seal settles down near other females. She needs to rest. For 11 months, a seal pup has been growing inside her. In two days, the new seal will arrive.

CLOSE-UP

True seal or sea lion?

True (or Common) Seal

- Front flippers cannot rotate backward; back flippers cannot move forward
- No external ears
- Short neck

Sea Lion

- Front flippers rotate backward; back flippers can move forward
- External ears
- Long neck

CHAPTER ONE

LIFE BEGINS

Seals are **marine mammals**. They are semiaquatic, spending part of their time in the sea and part on land or ice. Seals are shaped like torpedoes. Their bodies are designed to move effortlessly through the water. A thick layer of fat, called blubber, lies just below their skin. Blubber and fur help keep seals warm in cold water.

Seals belong to the pinniped family. Animals in this family have four fin-shaped flippers. There are three groups of pinnipeds: earless seals, eared seals, and walruses. Earless seals are also known as true seals or common seals. They really do have ears, but the tiny openings on the sides of their heads are small and barely noticeable.

Nineteen species of true seals live all over the globe in cold oceans, including the Pacific and Atlantic Oceans. The smallest adult seal, the Baikal seal, measures just over three feet (one meter) long and weighs up to 200 pounds (91 kilograms). An adult southern

NORTHERN ELEPHANT SEAL MILESTONES

DAY 1

- Born
- Black furry coat
- Eyes open
- Weight: 70 pounds (32 kg)
- Length: 4 feet (1 m)

FEATURED FAMILY

Welcome to the World

The female elephant seal rocks gently on her large belly and tosses her flippers back and forth. She lifts her head and moans. Soon, the seal pup's head appears. With another bark and another push, the newborn flops onto the sand. The mother seal mewls and grunts as she inspects her new baby. She touches the newborn nose-to-nose. The female pup wiggles on the sand, testing the motion of her flippers. The pup finds her mother with her snout before letting out a sharp bark. It is time to eat.

elephant seal can be 19 feet (6 m) in length and can weigh almost 9,000 pounds (4,100 kg).

Most seals feed on fish, squid, and crustaceans. Leopard seals, however, eat penguins, seabirds, and sometimes other seals. The majority of seals hunt and forage for food on their own. Some seals, like the harp seal, travel and hunt in large groups. Orcas, sharks, polar bears, wolves, and foxes all prey on seals.

Seals haul out on beaches or on pack ice to molt, breed, and deliver young. At these times, they often congregate in large numbers. Seals get quite vocal in groups. Males bark, grunt, and fight other males. Female seals and pups yelp and whine to add to the cacophony.

Male seals are called bulls. Female seals are called cows. With the exception of spotted seals, bulls do not help care for pups. The cows do all the work, but not for very long. Although newborn pups grow rapidly while nursing on their mothers' milk, most are weaned after only four weeks or less. The fatty milk helps thick layers of blubber form on the pups' bodies.

CLOSE-UP

Flippers

All seals have four flippers. The short front flippers are webbed with blunt claws at the end of each digit. The back flippers are also webbed with five bony digits. Moving the back flippers from side to side pushes the seal through the water.

1 WEEK

- Lives on mother's milk
- Stays close to mother on sandy beach
- Gains weight rapidly

2 WEEKS

- Recognizes mother's bark
- Continues to gain weight rapidly
- Moves about on front flippers

CLOSE-UP

Recognition

As soon as a seal pup is born, mother and baby learn to recognize each other. The mother touches her snout to the pup's so the baby can smell her. She yelps and barks often so that her pup will know her voice.

FEATURED FAMILY

First Meal

Moments after giving birth, the mother elephant seal moves close to her newborn and rolls over onto her side. She touches the pup with her front flipper and guides the baby to her body. The pup pushes her snout onto her mother's belly, finds a nipple, and begins drinking ultra-rich milk. In the early days of nursing, a mother seal's milk contains about 12 percent fat. By the time nursing is finished in about four weeks, the milk will be almost 60 percent fat. By comparison, human milk contains about five percent fat.

WHEN SEALS *haul* out of the water, they CONGREGATE in large NUMBERS.

3 WEEKS

- Flippers getting stronger
- Barks, grunts, and mewls to mimic mother

CLOSE-UP

Fur

About once a year, seals come up on land to shed their fur and outer layer of skin. This is called molting. During molting, seals go without food. Their furless bodies can't withstand hunting in the cold ocean water.

CHAPTER TWO

EARLY ADVENTURES

Once they wean their pups, seal mothers usually go back into the sea. They never return to their offspring. Females may stay on land long enough to molt, but they rarely interact with pups during that time.

Abandoned seal pups stay on land or ice, surviving on their fat reserves, until their hunger entices them to leave. Then, they slip into the water just like they've seen their mothers do. They quickly learn to swim and find food.

At first, pups search for small crustaceans in shallow waters. They are too young to dive very deep. Their abilities rapidly improve, though. Pups who don't master the skill of catching food struggle to survive.

4 WEEKS

- Nursing stops when mother goes back to ocean
- Abandoned by mother and left on beach with other seal pups
- Practices short swims and dives at water's edge
- Weighs close to 300 pounds (136 kg)

The special features of a seal's body help it find food. Large eyes let seals see in dark water. Specially shaped whiskers detect the movements of underwater prey. Keen hearing helps seals find food and avoid predators. A seal's sleek body allows it to move through the water quickly and smoothly to chase prey and flee from danger. Sharp, pointed teeth help seals grasp and hold onto their food until they can swallow it whole.

CLOSE-UP

Sight

A seal's large eyes can focus well underwater, even when the water is dark and murky. This helps them locate prey and avoid predators. However, on land, seals can clearly see only objects that are fairly close.

After months of practice, young seals are able to dive deep into the ocean to forage for food. As opportunistic feeders, seals aren't picky about a meal. They eat whatever fish, octopus, squid, or crustacean they can find.

FEATURED FAMILY

Look Who's Swimming

For four weeks, the newborn elephant seal nurses frequently. The mother seal does not eat at all. Then, abruptly, nursing ends. The mother seal wriggles back into the ocean, abandoning her pup forever. By now, the pup weighs around 300 pounds (136 kg). For the next two months, she stays alongside other abandoned seal pups near the beach. Unable to hunt food on her own, the weanling lives off body fat reserves. Then, one day, she cautiously approaches the water's edge. She splashes her flippers back and forth in the surf. She ducks her head underwater and makes a shallow dive. If the young seal wants to eat, she must learn to swim.

6 WEEKS

- Molts and turns light silver
- Dives into the water in search of food when fat reserves are depleted
- With practice, learns to catch prey

CLOSE-UP

Hearing

Although they are classified as earless seals, true seals or common seals do have small ear openings on either side of their heads. Seals hear well underwater. They are experts at determining where sounds come from, which is critical for locating prey.

FEATURED FAMILY

Give It a Try

Hunger forces the three-month-old seal to journey into the sea. She keeps to shallow waters. Using her back flippers, the young seal pushes her way, nose-first, to the ocean floor. She spots a shrimp and lunges toward it. The shrimp escapes, and the seal must swim back to the surface for a breath of air. On her next dive, the seal watches a lobster scoot under a rock. Darting through the water, the seal nudges the rock with her snout, slurps up the lobster, and swallows it whole. Success!

SPECIALLY SHAPED WHISKERS *detect* the movements of underwater PREY.

6 MONTHS

- Returns to rookery to molt
- Spends two months on beach, shedding fur and top layer of skin

8 MONTHS

- Begins first long ocean journey
- Makes impressively deep dives compared to other diving mammals

CLOSE-UP

Migration

Elephant seals migrate farther than any other mammal except for whales. Making two migrations each year, the elephant seal travels almost 12,000 miles (19,300 kilometers) in all. Males and females migrate to different locations to forage for food.

CHAPTER THREE

LIFE LESSONS

As they mature, seals become expert divers. Elephant seals dive almost a mile (1,525 m) below the surface to search for food on the ocean floor. They can stay underwater for two hours without coming up for air. Smaller seals, like harbor seals, dive 1,500 feet (457 m) and stay underwater for 30 minutes at a time. The ability to swim to extreme ocean depths often protects seals from sharks and orcas that want to prey on them.

Seals are specially equipped to withstand these deep dives. Layers of blubber keep them warm as the water temperature drops. Seals can slow down their heart rate to conserve energy and oxygen. A seal's normal heart rate is 80 to 120 beats per minute. When they are preparing to dive, seals reduce that rate to as low as three or four beats per minute. Seals can even sleep for a short time while submerged in the water.

15 MONTHS

- Returns to birthplace for annual molt

5 YEARS

- Females deliver their first pup

Adult seals travel on long migrations to search for food. Males and females usually go to different areas, sometimes traveling across vast stretches of the planet. Their migrations can be 6,000 miles (9,660 km) each way.

Each year, adult seals return to land and haul out long enough to molt. They shed their outer layer of fur to reveal a new layer beneath that is healthy and free from algae. After molting, seals return to the water for another migration. They spend eight or nine months of each year in the open sea.

Seals reach sexual maturity when they are three to six years of age. When it's time to give birth, pregnant cows return to familiar places on land or on **ice floes**. Male seals arrive at the same places several months earlier to prepare for mating.

FEATURED FAMILY

This Is How It's Done

The young elephant seal takes a breath at the ocean's surface before making another short dive. She flaps her strong flippers as she pushes toward the ocean floor. Suddenly, a squid propels its body through the water. The seal snaps at it! The squid darts away just as a hammerhead shark rushes toward the seal. Frightened, she dives deeper into dark, murky water. She stays concealed in the gloom until she needs air. The seal surfaces for a quick breath, then dives deep into safer waters.

CLOSE-UP

Hydration

Unlike humans and most other animals, seals do not need to drink water to stay hydrated. Despite being surrounded by water, seals get the water they need from the foods they eat.

7 YEARS

- Proboscis, or snout, is fully developed
- Males weigh up to 4,500 pounds (2,041 kg)
- Males measure 13 feet (4 m) long or more
- Females weigh up to 1,300 pounds (590 kg)
- Females measure up to 10 feet (3 m) long

9 YEARS

- Males use proboscis to vocalize threats to other males
- During breeding season, males form a collection of females to mate with
- Older females secure the best, safest locations for their pups

Some bulls fight and push each other around to establish a **hierarchy**. Only the dominant males will get the chance to mate with cows. Scientists think that repeated fighting among males may reduce their lifespan. Depending on their species, seals live for 13 to 30 years. Females tend to live longer than males.

Each cow delivers a single pup. Once the newborn pups are weaned, dominant males mate with the females. The adults return to the ocean for the next migration. Months later, the next generation of seals will be born.

CLOSE-UP

Camouflage

Seal pups left on ice floes are vulnerable to predators like polar bears, wolves, and foxes. Adaptive coloration makes them difficult to see. Harp seal pups are covered with fluffy white fur. This helps them blend in with the snow.

FEATURED FAMILY

Practice Makes Perfect

The young elephant seal has grown bigger, stronger, and more skilled. When her specialized whiskers sense movement in the water, she reacts quickly to locate the source. The seal chomps on a bluefish and holds it tight with her sharp front teeth. Slinging her head back, she swallows the fish whole. She dives again into dark water and uses her keen sense of sight to hunt octopus and other prey. The whole time, she is on the lookout for predators such as sharks and orcas. Now an expert diver, she can wait longer between breaths and forage on the ocean floor for longer and longer stretches of time.

DEEP
ocean **divers**

GREAT WHITE SHARK: 4,000 feet/1,200 m

NORTHERN ELEPHANT SEAL: 5,000 feet/1,500 m

CUVIER'S BEAKED WHALE: 9,500 feet/2,900 m

(13) **YEARS**

End of life for males

(19) **YEARS**

End of life for females

CHAPTER FOUR

HELPING SEALS SURVIVE

Throughout history, humans have killed seals for their meat, blubber, and furry **pelts**. For centuries, there were no laws about how many seals could be hunted. Their populations shrank.

In 1969, Brian Davies founded the IFAW (International Fund for Animal Welfare) to end the killing of seals in Canada. The organization grew into a worldwide effort to rescue, rehabilitate, and release animals. Seal hunting slowed down, but it didn't stop. The Caribbean monk seal disappeared from the planet.

The United States passed a law called the U.S. Marine Mammal Protection Act of 1972. It made it illegal to kill, take, or bother marine mammals in the United States. Since 1972, no marine mammal species have gone extinct.

Other countries have banned or limited seal hunting. Some have made it illegal to buy products made from seals. Seals have recovered in great numbers, but Mediterranean seals and Hawaiian monk seals are endangered.

Even though many seal populations around the world are stable, pinnipeds continue to face threats. Their food sources are shrinking. They often get entangled in fishing gear. Humans have polluted many seal habitats. All of this makes seals more likely to get diseases.

Climate change is the greatest threat to seals, especially seals that depend on pack ice and ice floes. When ice melts too early in the season, pups may become stranded and die. When food sources move to colder waters, seals can starve.

Conservation efforts around the world help protect seals. SeaWorld San Diego rescues seals that get stuck on California beaches. The Marine Mammal Center in Sausalito, California, is the world's largest marine mammal hospital. It cares for sick and injured animals and educates visitors about the threats faced by seals.

Scientists are learning new things about seals. A recent study showed that elephant seals sleep only two hours a day while in the water. Staying awake longer helps them avoid predators.

Knowing more about the lives of seals helps people find ways to protect them. With continued efforts, people can help seals and their four-flippered relatives survive.

OCEAN

FAMILY ALBUM

SNAPSHOTS

Pacific harbor seals can swim at birth, but pups sometimes ride on their mothers' backs when they get tired.

With only 1,500 left, **Hawaiian monk seal**s are endangered. Their Hawaiian name, 'Ilio holo I ka uaua, means "dog that runs in rough water."

Harp seals have a patch of black fur on their backs that resembles a harp.

Despite their name, **fur seals** are not true seals because they have external ears. These animals spend their summers off the coast of Alaska.

Gray seals live in coastal waters of the North Atlantic Ocean. Adults have large curved noses. Males have heads shaped like horse heads.

Weddell seals are one of the few species of seals that can give birth to twins.

Ribbon seals have alternating bands of dark and white fur. They live in Alaska and Russia.

Bearded seals get their name from the long white whiskers on their faces. They tend to stay near pack ice or ice floes.

Male **hooded seals** have a stretchy hood inside their noses that inflates to look like a red balloon. The inflated hoods attract females and threaten other males.

Unlike most true seals, **spotted seals** form family groups during the breeding season. Each group has a male, a female, and a pup.

The only seal to live in fresh water, **Baikal seals** are the smallest true seals. They are found only in Lake Baikal in the Russian province of Siberia.

Surprisingly, **crabeater seals**, found on the ice of Antarctica, do not eat crabs. Their diet consists of tiny crustaceans called krill.

The largest of all seals, **southern elephant seals** live in the cold waters of the Antarctic Ocean.

WORDS to Know

cacophony a loud, unpleasant mixture of sounds

crustaceans aquatic animals with jointed legs, such as lobsters, crabs, and shrimp

hierarchy a ranking according to status or dominance

ice floes ice sheets that float on the ocean's surface

marine mammals warm-blooded animals that depend on oceans for survival; mammals have backbones, produce milk to feed their young, and have hair or fur

molt to shed the outer layer of fur, skin, or feathers to enable new growth

opportunistic taking advantage of whatever opportunities or resources are available

pelts animal skins with fur still attached

rookery an area where a large number of animals or birds congregate to breed, nest, give birth, and raise young

weaned no longer getting food by nursing from a mother animal; at a stage of development in which an animal must feed itself

LEARN MORE

Books

Goodridge, Harry, and Lew Dietz. *A Seal Called Andre*. Rockport, Maine: Down East Books, 2014.

Klimp, Nancy Oline. *Meet Me at the Salish Sea*. London: MCP Books, 2021.

McKnight, Diane. *The Lost Seal (Long-Term Ecological Research)*. Boulder, Colo.: Taylor Trade Publishing, 2016.

Websites

"Seals." National Geographic. https://www.nationalgeographic.com/animals/mammals/facts/seals-pinnipeds-walruses-sea-lions

"Seals & Sea Lions." NOAA Fisheries. https://www.fisheries.noaa.gov/seals-sea-lions

"Seals, Sea Lions and Walruses." The Marine Mammal Center. https://www.marinemammalcenter.org/animal-care/learn-about-marine-mammals/pinnipeds

Documentaries

Attenborough, David. *Male Elephant Seals Defend Territory*. BBC Wildlife, 2007.

Harbor Seal Monitoring in Puget Sound. NOAA, 2013.

Wharton, Drew. *A Seal's Life: The Story of the Northern Elephant Seal*. Wharton Media, 2007.

Note: Every effort has been made to ensure that any websites listed above were active at the time of publication. However, because of the nature of the Internet, it is impossible to guarantee that these sites will remain active indefinitely or that their contents will not be altered.

Visit

GEORGIA AQUARIUM

Take a behind-the-scenes tour of the seal habitat and learn from experts.

225 Baker St NW
Atlanta, GA 30313

MARINE MAMMAL CENTER

Visit the world's largest marine mammal hospital to see animal care providers taking care of seals and other marine animals. Check out their website to adopt a seal patient.

2000 Bunker Road
Sausalito, CA 94965

NEW ENGLAND AQUARIUM

Watch a daily training session with Atlantic harbor seals.

1 Central Wharf
Boston, MA 02110

POINT REYES NATIONAL SEASHORE

Take the Elephant Seal Overlook Walk to gaze down at hundreds of seals on the beach.

1 Bear Valley Road 70
Point Reyes Station, CA 94956

INDEX